Chemical Engineering Made Simple

Process to Progress

Diana Tran

Copyright © 2021   By Diana Tran

All rights reserved.

No part of this publication may be reproduced or transmitted in any form or by any means, electronic or mechanical, including photocopy, recording, or any information storage or retrieval system, without permission in writing from the copyright owner.

For intended purposes, the illustrations have been simplified to understand a **few key concepts** related to the chemical process. It is **by no means** a comprehensive representation and/or replacement of the actual processes.

Designed by
theillustrators.com.au

To my brother for being there every step of the way.

To dear friends and colleagues for your support and encouragement.

What is chemical engineering?

Essentially, it involves taking a **RAW MATERIAL**, putting it through a **PROCESS** to create a **PRODUCT**, and then with time **PROGRESSION** occurs. In the first instance, it can easily be explained as a 4-stage pictogram.

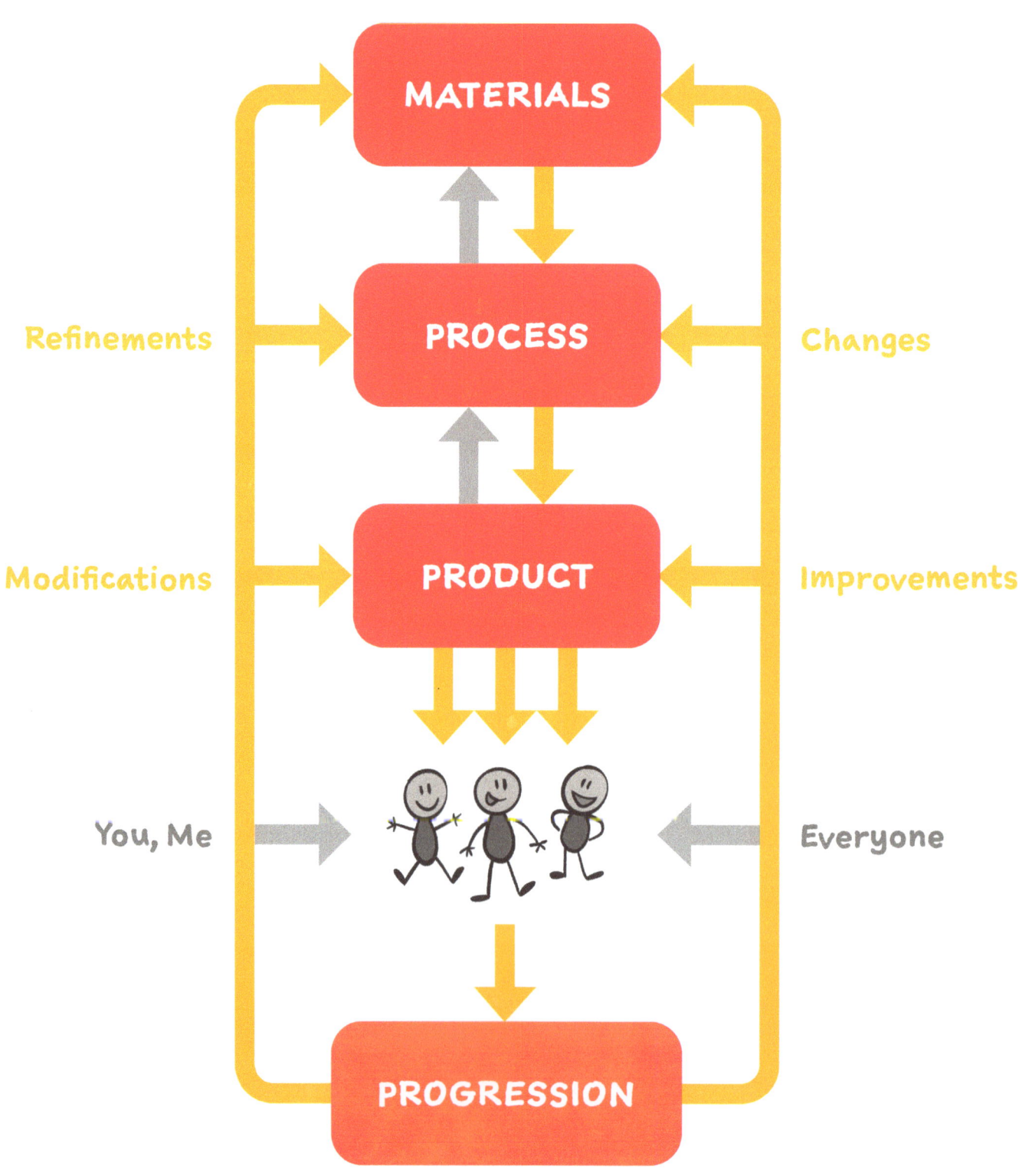

Still struggling?

Let's take an example, baking a sponge cake.

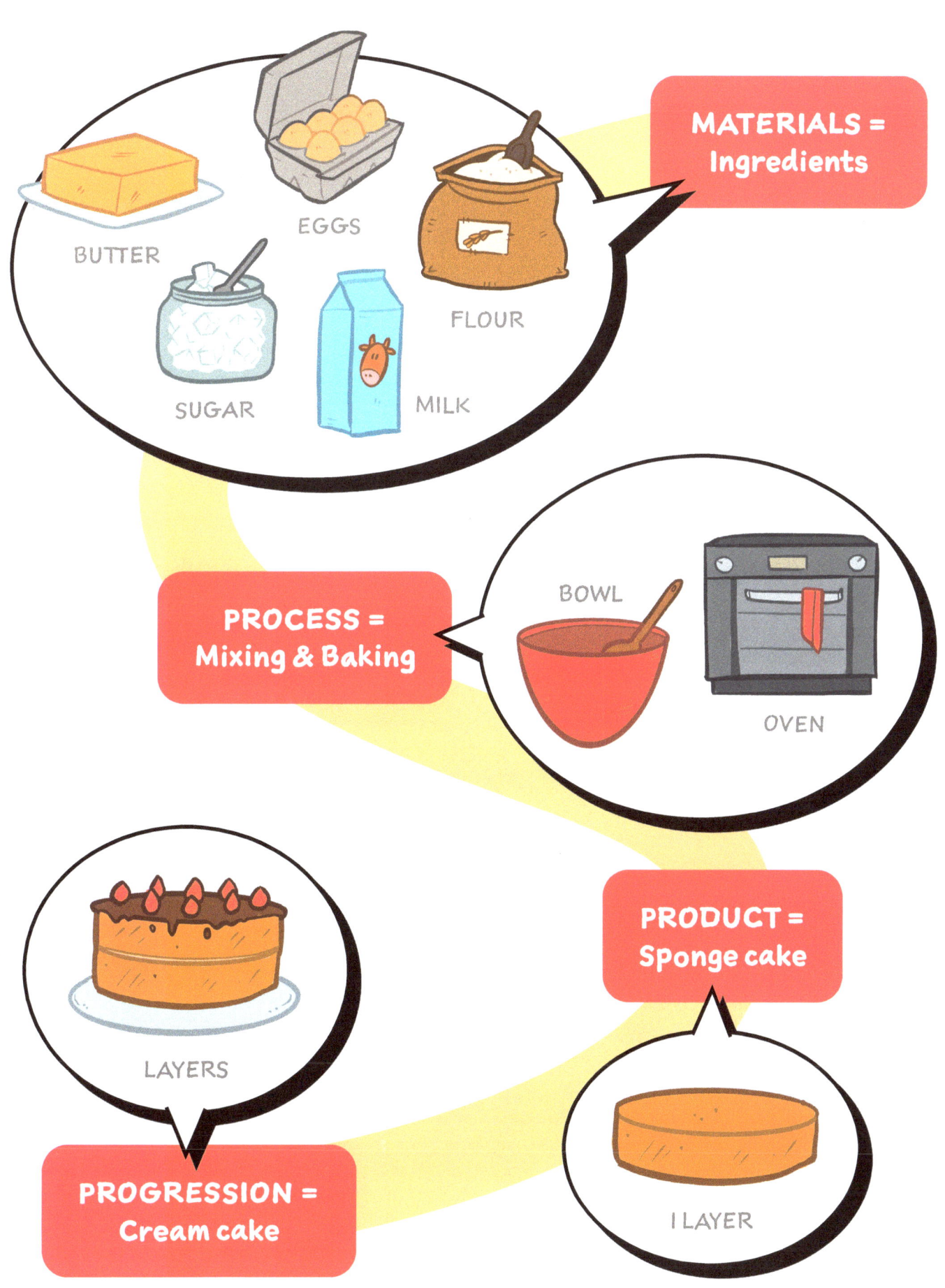

So back to chemical engineering...
What is it?

It's in the food we eat, the beverages we drink, the clothes we wear, the electricity we use, ... it is everywhere!

Let's break down some of the common things in everyday life and see the chemical process behind it as a 4-stage pictogram and through the simple eyes of me.

What goes into making a coffee?

The coffee beans are broken down into smaller particles and then packed inside a column. The flavour of the coffee depends on how long the boiling water is in contact with the particles.

This is known as the *brewing* process.

Coffee

1. MATERIALS

Coffee beans

> * **Packed bed column:** a column filled with a packing material to allow a liquid to flow down through it; increases the contact time between the liquid and material.
>
> ** **Solid-liquid extraction:** a chemical process for removing a soluble component from a solid by using a liquid.

2. PROCESS

- Grinding
- Packing
- Extracting

* **Packed bed column**

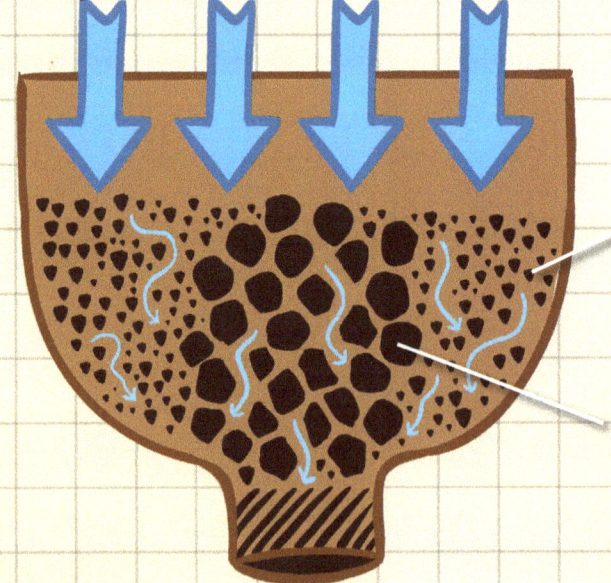

Water flow, Q (cm/s)

fine

coarse

** **Solid-liquid extraction**

3. PRODUCT

4. PROGRESSION

In one's own home

Brew Time, t (s)
Short: Under (not enough)
Long: Over-powering (bitter)

How does milk change into a powder?

The liquid milk is passed through a nozzle attached to a glass chamber where it changes into tiny droplets. The hot air helps to carry the droplets through as it dries into a powder. However, if the temperature is not correct the droplets will end up somewhere else!

To do this, a *spray drying* process is used.

....

Spray dryer image adapted from BUCHI Spray Dryer & Encapsulation Solutions, Particle formation for lab scale Brochure, www.buchi.com/spray-drying/solutions.

Milk Powder

1. MATERIALS

Dairy cow (milk)

> * **Spray dryer:** an instrument used to convert a liquid solution into a dry powder through a fast drying process. The process can depend on the spray air flow and the inlet (T_{inlet}) and outlet temperature (T_{outlet}).
>
> ** **Nozzle:** used to apply pressure to the liquid to change its form into a fine mist.

2. PROCESS

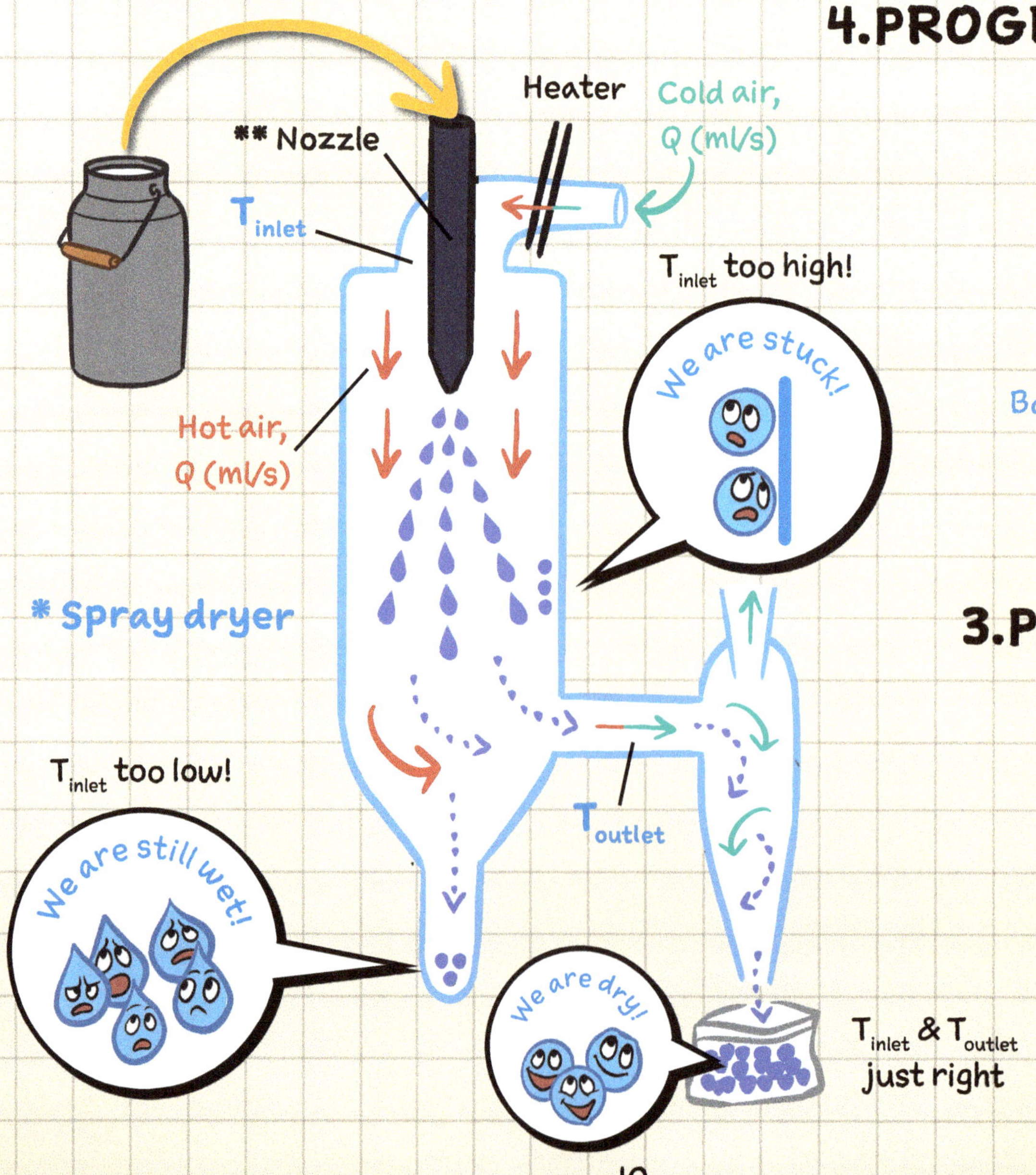

- ** Nozzle
- Heater
- Cold air, Q (ml/s)
- T_{inlet}
- Hot air, Q (ml/s)
- * Spray dryer
- T_{outlet}
- T_{inlet} too low! — We are still wet!
- T_{inlet} too high! — We are stuck!
- We are dry! — T_{inlet} & T_{outlet} just right

3. PRODUCT

MILK POW

4. PROGRESSION

Baby formulation

Why does perfume smell nice?

The floral is mixed with a liquid, which helps to release its essential oils. The oils are carried in the steam where it is changed back into a liquid. The liquid is then passed through a separator to remove the essential oils from the excess water. The essential oils are what we smell. How would you do this?

By going through a *distillation* process.

Perfume

1. MATERIALS

Floral

> * **Distillation column:** a column used to extract the liquid from its component parts, based on how readily it vapourises. The steam is then converted to a liquid via a cooling process.
>
> ** **Separator:** separates the different components based on their density (oil is lighter, water is heavier).

2. PROCESS

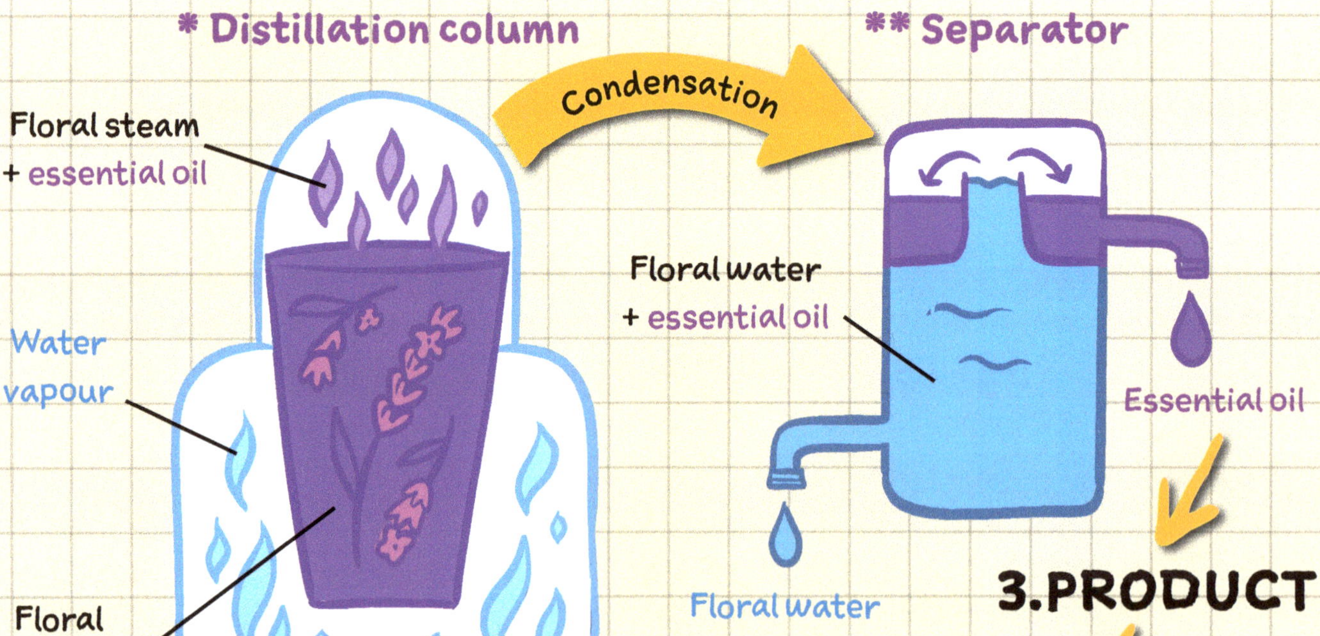

* Distillation column → Condensation → ** Separator

- Floral steam + essential oil
- Water vapour
- Floral + liquid
- Water
- Heat

- Floral water + essential oil
- Essential oil
- Floral water

3. PRODUCT

4. PROGRESSION

Parfum (female)   Cologne (male)

How is sunscreen made?

There are a few ingredients (each with its own function) that are mixed together rapidly to form a cream. The cream contains many stable small droplets. One of the chemical substances used to protect us from the sun's rays is called zinc oxide (ZnO).

This is known as a *batch mixing* process.

High shear mixer image adapted from Silverson Mixer Operation Manual, www.silverson.com.

Sunscreen

1. MATERIALS

Ingredients
- **Chemicals** — active solids
- **Stabilisers** — help bind the product together
- **UV additives** — protects from the sun (ZnO)

* **High shear mixer:** a mixer that uses the action of shearing to combine solid powders that naturally cannot mix together into a liquid to form a cream.

** **Rotor head:** creates a strong suction to draw the solids and liquid together where they are rapidly mixed.

*** **Water-in-oil-emulsions:** small droplets that contain the solid ingredients in the water inside an oil coating.

2. PROCESS

* **High shear mixer**

2000 rpm | 1h 15s on 30s off

Ingredients + water

Mixing Time, t (h)
- Short: Not stable
- Long: Overheating; water loss, phase separation

→ Oil / Water (separated)

Loss of water!
We are stuck!

** **Rotor head** — Stable and constant

*** **Water-in-oil emulsions**

3. PRODUCT

SPF = Sun Protection Factor

(SUN SCREEN)

4. PROGRESSION

SPF 15+ → SPF 30+ → SPF 50+

14

How does a tablet form its shape?

The powders that have been mixed together and dried are fed in between two halves of a mould. The mould comes together to compress the powder into shape. Whether the tablet holds its shape is determined by the pressing speed and the applied pressure.

This is called the *compression* process.

Tablet

1. MATERIALS

Active Ingredients
Dissolvable powders and drug components

* **Tablet press:** a device that compresses powder into a uniform size tablet of equal weight.

** **Compression:** used to reduce the volume by applying pressure.

*** **Die:** a mould that gives the tablet its shape when the powder is compressed (located in the punch area).

2. PROCESS

- Mixing
- Drying
- Pressing

* **Tablet Press (punch area)**

Compression rollers

Upper punch

*** Die

Lower punch

Punches move at the same time

Speed (mm/s) is important!
Punches come together
Powder
Punches release

** **Compression**
Loose: pill weak and crumbles
Too tight: pill takes too long to dissolve in the body

3. PRODUCT

4. PROGRESSION

Variety

How is a fertiliser created?

The nutrients are blended together in a rotating drum where binders and liquids can be added. The rotational speed of the drum generates heat that causing a chemical reaction to occur within the materials. Hundreds of small spherical solid pellets can be formed in the drum!

We call this the granulation process.

Fertiliser

1. MATERIALS

Nutrients
- Nitrogen — N^+
- Phosphate — PO^{4-}
- Hydrogen — H^+

Binders

*** Granulation drum:** a rotating drum used to tumble materials in the presence of a liquid to produce spherical solid particles.

**** Centrifugal force:** an outward force which acts on a moving body in a circular motion. This force can overcome gravity if it is too high.

2. PROCESS

Rotational Speed, v (m/s)
Too slow: powders stay at the bottom
Too fast: powders stay at the wall

** Centrifugal force

4. PROGRESSION

Plants thrive

B^+
K^+
Mg^{2+}
Cu^{2+}

Gravity — Spreading — Movement of powder — * Granulation drum — solid pellets — water

3. PRODUCT

Why is a potato chip crispy?

The slices of potatoes are deep-fried in a large quantity of oil. The high heat in the oil is perfectly controlled to give the potato its crunch. The quality of the oil also adds texture and flavour to the chips which is further enhanced with seasoning.

Let's take a look at the deep-frying process.

Potato Chip

1. MATERIALS

Potatoes

> * **Deep-fryer:** a device used to cook food in oil at high temperatures.
>
> ** **Oil mixture:** oil with small amounts of water from the potato. Oil is lighter than water and remains at the top, while water falls to the bottom to be removed.

2. PROCESS

- Peeling
- Frying
- Flavouring

Slices of potatoes (off conveyor belt)

** **Oil mixture** (lighter than water)

* **Deep-fryer**

Water is separated (heavier than oil)

3. PRODUCT

Heat and Time
Not enough: undercooked
Too much: overcooked/burnt

4. PROGRESSION

Air fryer for home

How is a candle made?

The wax is melted to a liquid which is then poured into moulds. The moulds contain the wick. When the liquid has cooled down and harden the wick is cut to remove the candle. Blowing out candles is a must for one's birthday!

We called this the wax melting process.

Candle

1. MATERIALS

Beeswax

* Wax melting: pieces of scented wax that are gently heated to form a liquid that can be poured. Natural waxes can be harvested from bee hives and synthetic waxes are known as paraffin.

** Wick: normally made of a braided cotton to hold the flame of the candle and influences how the candle burns.

2. PROCESS

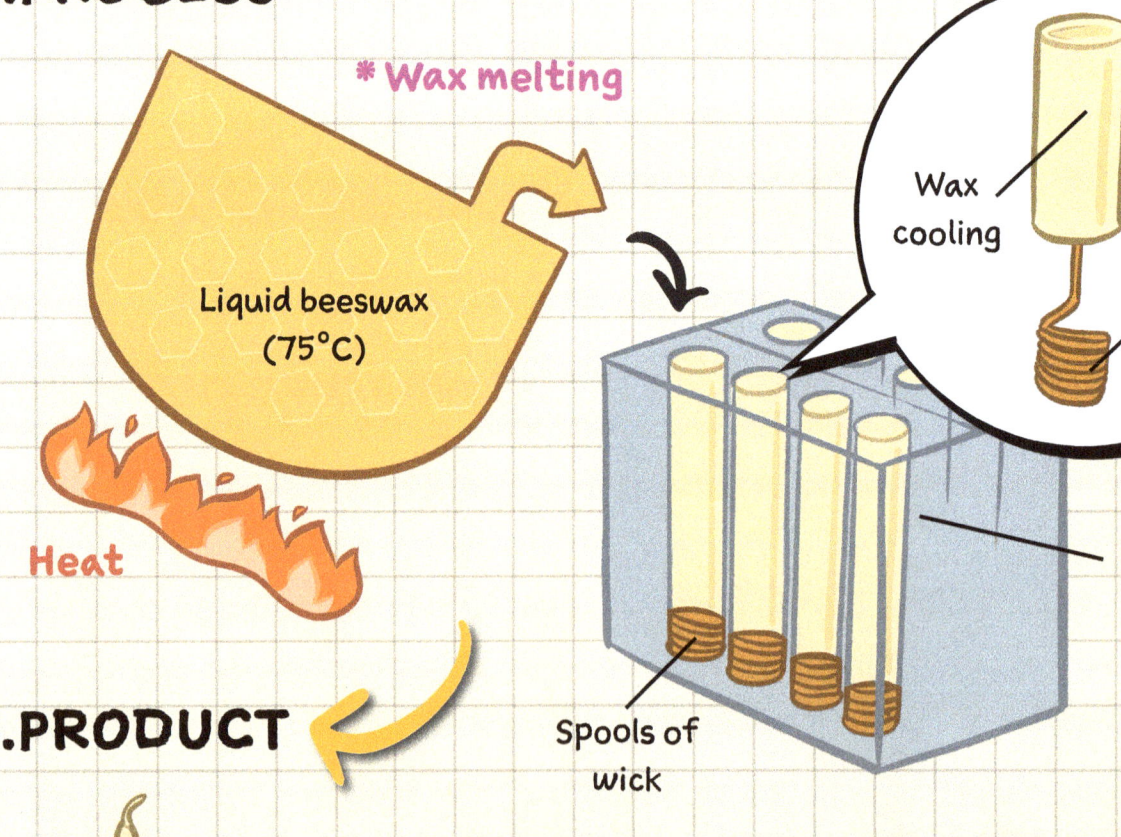

3. PRODUCT

** Wick
Pillar candle: burns longer
Tea light candles: burns shorter

4. PROGRESSION

Why are demin jeans indigo?

The fabrics are dipped in large tanks containing the vat dye. The insoluble dye is reduced in a base bath so that it can stick to the fabric fibres. The shade of the coloured jeans depends on the number of dipping cycles.

This is known as the vat dyeing process.

Jeans

1. MATERIALS

Fabric

> * **Vat dye:** a water-insoluble dye that is applied to a fabric in a reducing bath to make it soluble. The colour is obtained through an oxidation process in the fabric fibres.
>
> ** **NaOH:** sodium hydroxide - a chemical substance used to reduce the dye by changing the bath solution to pH 12-14.

2. PROCESS

- Reduced form = yellow
- Oxidised form = light blue
- Each dip = colour darkens

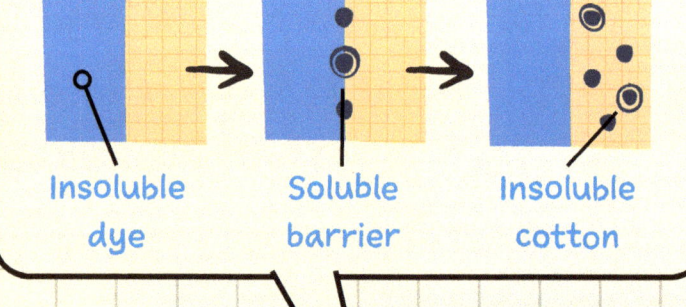
Insoluble dye → Soluble barrier → Insoluble cotton

* Vat dye + ** NaOH

Cotton

Heat (60°C)

3. PRODUCT

4. PROGRESSION

How is soap made?

It is the result of a chemical reaction between an acid (fat/oil) and a base (salt). The fats are heated to 40°C before a liquid base is added to react and thicken the mixture. The mixture is then poured into moulds and allowed to set, which can take hours to days!

We call this the *saponification* process.

Soap

1. MATERIALS

Fats or oils

Lye = sodium hydroxide

> *** Saponification:** a chemical reaction to convert fat using a strong alkaline to produce soap.
>
> **** CSTR (continuous stirred tank reactor):** a tank of constant volume with a stirring system that provides efficient mixing. The output of the batch process can be determined with accuracy and precision.

2. PROCESS

* Saponification

- Motor
- 250 ml/min
- 30 min
- Fat + Lye
- ** CSTR
- Heat (40°C)

Washes away germs

4. PROGRESSION

Moulds

3. PRODUCT

26

How is a light bulb shaped?

Clear liquid glass is produced by melting sand in a furnace above 1000°C. A long metal rod is used to pick up a bob of molten glass, which expands when air is blown through it. Rotating the rod helps to shape the glass.

Let's take a look at the glassblowing process.

Light Bulb

1. MATERIALS

Sand (silicon oxide)

* **Glassblowing:** a glass forming technique that involves shaping molten glass into a bubble by blowing air into it via a long metal tube.

** **Melting:** a process for melting sandy material in a furnace at 1200 °C to form clear glass.

2. PROCESS

Mixing

** Melting

*Glassblowing

Metal tube

Furnace

Air → Expand and form

3. PRODUCT

Working the Glass (timing is key!)

Too long: hardens before shaping occurs

Not enough: imperfections in the product

4. PROGRESSION

28

What is inside a light bulb?

The active component known as tungsten is extracted from an ore, which is crushed down and processed through a series of chemical reactions to produce a powder. The powder is then mixed with a binder to form a paste that is converted to pure tungsten when heated under a hydrogen atmosphere.

A *thermal annealing (heating)* process is used.

Filament

1. MATERIALS

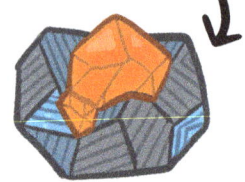

Combined with ** Chemicals

* **Scheelite:** a calcium tungstate mineral with chemical formula $CaWO_4$.

** **Chemicals:** sodium hydroxide (NaOH), calcium carbonate ($CaCO_3$), hydrochloric acid (HCl), ammonia (NH_3), polymer binder

*** **Thermal annealing:** a heat treatment that is used to alter the materials physical and/or chemical properties with temperature and time. Commonly used to strengthen the metal.

2. PROCESS

- Crushing
- Chemical reactions

*** Thermal annealing

Hydrogen gas — Hydrogen gas

Furnace (>600°)

Gas
Carbon: tungsten carbide (wrong form)
Hydrogen: PURE tungsten (correct form)

3. PRODUCT

Tungsten (W)
(strongest metal in the known universe)

4. PROGRESSION

What is chemical engineering?

... it is about working with materials ...

... evaluating the time it takes through a process ...

... producing efficient products to make our lives easier...

... that progresses the world that we see and live in today ...

... it is everywhere!

Where else can you see this?

www.ingramcontent.com/pod-product-compliance
Lightning Source LLC
Chambersburg PA
CBHW041712290426

44109CB00028B/2851